Contents

The material for years 3, 4 and 5 of SMP 11–16

The yellow, blue, red and green series together make up the second part of the SMP 11–16 course, for pupils in the third to fifth years of secondary school (ages 13+ to 16+).

(The booklet scheme which forms the first part of the course is fully described in the *Teacher's guide* for levels 1, 2, 3 and 4, and the *Practical guide*.)

The overall structure of the material for years 3 to 5 is set out in the diagram below.

The Y series is for the most able group of pupils (roughly speaking, the top 20% to 25% or so, although the proportion is likely to vary from school to school). The B and R series are for the 'middle' group (the next 35% to 40% or so) and the G series for lower ability pupils (apart from those with special learning difficulties).

The B series branches after *Book B2* to allow the more able of the pupils in the middle group to move ahead on to more demanding work in the R series. The mathematical content of the R series has much in common with that of *Books Y1*, *Y2* and *Y3*, but presentation and pace are often different.

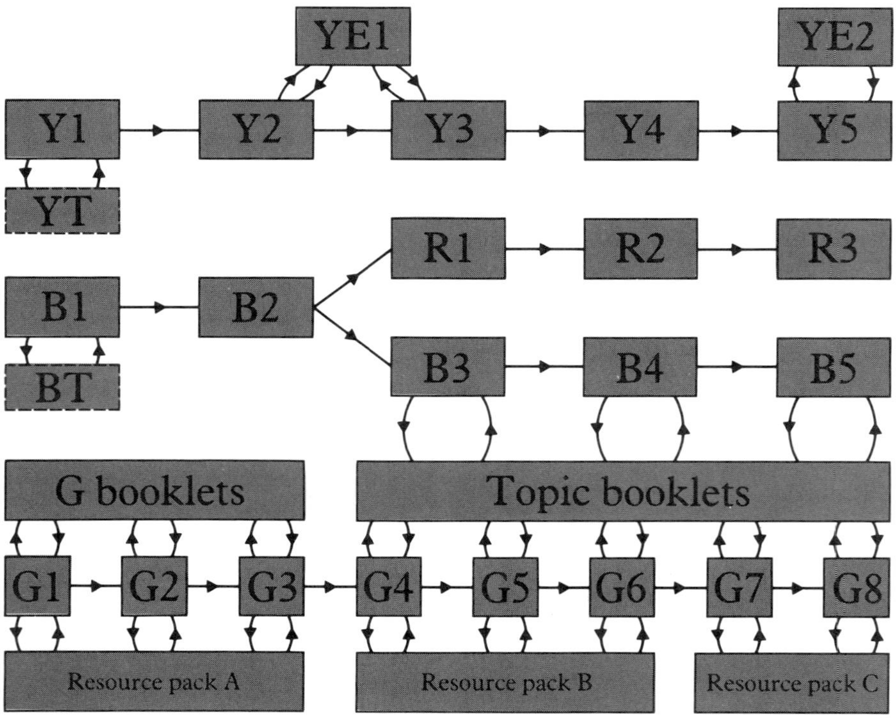

Classroom organisation and teaching style

Although there is rather more exposition and explanation in the books than is found in many other textbooks, the books are not intended to be 'self-instructional'. Many important points arise in the course of doing the problems in the books, and these points will need to be brought out by the teacher in discussion with the whole class or with smaller groups or as appropriate. Teachers may find it possible from time to time to give particular chapters or sections of chapters to the class to work through on their own, which is no bad thing since the ability to pick up information from the printed page and to follow written explanations is an important one. Where this is done it will be necessary for the teacher carefully to 'go over' what has been done.

Introduction to Book B5

Mental and written arithmetic and the use of calculators

It is assumed throughout that unless there is an instruction to the contrary calculators will be used for all but the simplest calculations which can be done mentally.

We strongly recommend that teachers encourage mental calculation and from time to time give short sets of questions to be answered mentally. We also suggest having occasional practice sessions on written arithmetic, but that the scope of these should not extend beyond addition, subtraction, multiplication by 2, 3, ..., 9 and division by 2, 3, ..., 9 of whole numbers and money.

Starred questions

Occasional questions are starred to indicate that they are of greater difficulty, and can be left out by slower pupils using the book.

Other material suitable for pupils using Book B5

The three series of topic booklets are written for pupils following either the G or the B series. A further source of material is to be found in the G resource packs.

For pupils who have not met them before, the investigations in the pack entitled *Investigations and stretchers* may be used in addition to the investigations included in the B books themselves.

Equipment needed

Certain standard items of equipment are needed frequently and no special attention is drawn to them in the books. These include rulers, compasses, scissors, and 2 mm graph paper.

In other cases, equipment needed is referred to in the book.
There are no worksheets needed for *Book B5*.

Pupils working from the B series are assumed to have the use of an ordinary four-function calculator.

Ordering equipment

The following items required for *Book B5* are published by Cambridge University Press. You should order them through your usual school book supplier.
Angle measurers (pack of 5) ISBN 0 521 25435 3
Pie chart scales (pack of 10) ISBN 0 521 26366 2

When ordering, remember to state the ISBN, the series title (SMP 11–16), the name of the item, the publisher and, where appropriate, the number of packs you want.

Notes and answers for Book B5

1 Speed

This chapter revises the relationships between speed, distance and time.

A Distance, speed and time

A1 210 miles

A2 75 miles

A3 2700 km

A4 40 m

A5 2·7 m

A6 (a) 105 m.p.h. (b) 170 km/h (c) 47 m/s

A7 (a) About 323 km/h (b) About 89 m/s

A8 About 13 m/s. Judith travels about 130 metres in 10 seconds.

A9 9 m.p.h.

A10 8 m.p.h.

A11 18 m.p.h.

A12 (a) (i) 49 m.p.h. (ii) 46 m.p.h.
(b) The Exeter–Manchester coach is faster on average.

B Calculating journey times

B1 5 hours

B2 (a) 4 hours (b) 7 hours (c) 2 hours

B3 $9\frac{1}{2}$ hours

B4 $14\frac{1}{4}$ hours (to the nearest $\frac{1}{4}$ hour)

B5 (a) (i) 8 hours (ii) 6 hours
(iii) 4·8 hours (iv) 4 hours

(b)

Speed in m.p.h.	24	12	8	6	4·8	4
Journey time in hours	10	20	30	40	50	60

(c)

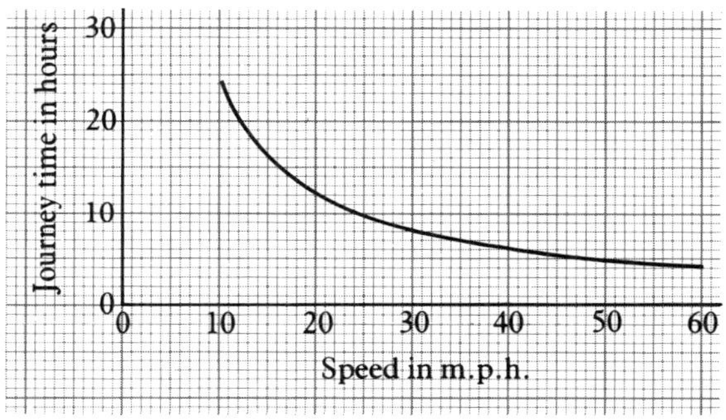

C Decimals of an hour (1)

C1 (a) 2 hours 30 minutes (b) 5 hours 43 minutes
(c) 3 hours 2 minutes (d) 15 hours 51 minutes

C2 (a) 3·29 hours (b) 3 hours 17 minutes

C3 (a) 2 hours 38 minutes (b) 3 hours 5 minutes
(c) 43 minutes (d) 26 minutes

C4 (a) 4 hours 25 minutes (b) 10:55 a.m.

C5 (a) 2 hours 43 minutes (b) 5:58 p.m.

C6 (a) 5 hours 21 minutes (b) 8:06 p.m.

C7 7:21 p.m.

D Decimals of an hour (2)

D1 (a) 4·27 hours (b) 3·12 hours
(c) 8·83 hours (d) 4·78 hours
(e) 0·22 hour (f) 0·47 hour
(g) 0·85 hour

D2 46 m.p.h.

D3 (a) 42 m.p.h. (b) 9 m.p.h. (c) 6 m.p.h.

D4 (a) 2 hours 35 minutes (b) 46 m.p.h.

E Mixed questions

E1 Aircraft A 481 m.p.h.
Aircraft B 470 m.p.h.
(both to nearest 1 m.p.h.)
So aircraft A flew faster on average.

E2 (a) 4·8 hours (b) 4 hours 48 minutes

E3 650 miles

E4 (a) 23 minutes (b) 0·38 hour (to 2 d.p.)
(c) 76 m.p.h.
(d) (i) 70 m.p.h. (ii) 73 m.p.h.
(iii) 74 m.p.h. (iv) 96 m.p.h.

Money matters: income tax

The six sections on money matters each provide a brief introduction to a financial topic. Further work on these topics is best done using up-to-date real information from tax offices, banks, building societies, etc., preferably getting the pupils to collect the data themselves. Some topics will be closer to their own interests than others.

Although the topic of income tax is generally included in mathematics courses because of the calculations involved, calculations which will be relevant to most pupils later in their lives, there are wider political issues which may already interest some pupils, and which arguably ought to be raised and discussed as part of the general education of every future citizen. (Why, for example, has it generally been accepted by all political parties that tax rates should increase as income increases, as shown in the diagram on page 11?)

2 Value for money

This chapter looks at comparisons which may be made when judging value for money. We can compare packages of different sizes by comparing either unit costs (cost per kg, per litre, etc.) or by comparing the quantities bought per penny or per £.

The overall level of prices in two shops can be compared by using a standard 'basket' of goods as the basis for the comparison. (This method also enables prices to be compared over a period of time.)

A Cost per kilogram, per litre, etc.

A1 (a) 19p per kg (b) 75p per kg
(c) £2·21 per kg

A2 (a) £3·60 per litre (b) £2·50 per litre
(c) £1·40 per litre

A3 (a) $13{\cdot}87\,\text{m}^2$ (b) £7·14 per m^2

B Comparing prices (1)

B1 (a) 44p per kg (b) 42·7p per kg
(c) The 3 kg bag is better value.
(d) Some people may not need 3 kg of potatoes.

B2 1·5 litre bottle – £2·23 per litre
2·5 litre bottle – £2·30 per litre
(to nearest p)

B3 1·8 m length – £1·42 per metre
2·4 m length – £1·35 per metre
(to nearest p)

B4 (a)

Surf	**Bold**
A 86·2p per kg	D 90·3p per kg
B 63·6p per kg	E 83·5p per kg
C 63·5p per kg	F 64·3p per kg
	G 72·4p per kg

(to nearest 0·1p)
(b) Pack C (c) Pack F

B5 (a) £8·25 per m^2 (b) £7·76 per m^2
(c) £8·33 per m^2
Carpet (b) has the lowest unit cost.

C Comparing prices (2)

C1 (a) 213 g tin – 12·5 g per p
439 g tin – 15·7 g per p
624 g tin – 13·0 g per p
(to nearest 0·1 g)
So the 439 g tin gives the best value for money.
(b) 213 g tin – 0·080p per g
439 g tin – 0·064p per g
624 g tin – 0·077p per g
(to nearest 0·001p)
So this confirms that the 439 g tin gives the best value for money.

C2 (a)

Own brand	**p per g**	**g per p**
220 g	0·055	18·3
439 g	0·034	29·3
539 g	0·046	21·6
Heinz		
150 g	0·087	11·5
225 g	0·067	15·0
450 g	0·044	22·5
580 g	0·047	21·5
840 g	0·045	22·1
2·62 kg	0·052	19·4

(b) £1·20 (c) 2·7 kg
(d) Six 450 g tins contain more than one catering tin, and cost less.

C3 (a) 6·8 pint can – 49·3p per pint
3·9 pint can – 45·9p per pint
The 3·9 pint can is better value for money.
(b) (i) 1 pint (ii) 23p (iii) 23p per pint
(c) Both cans might not be needed at the same time.

C4 (a) 330 ml can – 51·5p per litre
500 ml can – 52·0p per litre
1·5 litre bottle – 39·3p per litre
2 litre bottle – 37·5p per litre
The 2 litre bottle gives the best value for money.
(b) They may not drink a lot of it.

C5 (a) £1·15 (b) £1·08 (c) £1·20
(to nearest penny)
(d) 300 g, 200 g, 100 g, 500 g, 50 g
(e) three 300 g tins and one 100 g jar

C6 (a) 6 lots of 25 g; 7·5p per 25 g
(b) 250 g – 6·5p per 25 g
6 × 25 g – 9·5p per 25 g
9 × 25 g – 8·8p per 25 g
(to nearest 0·1p)
(c) They can be opened when they are needed.

D Comparing prices in different shops

D1 Supermarket A – £6·03
Supermarket B – £5·68
Supermarket C – £5·60

D2 (a) A – £8·54; B – £8·04; C – £7·39
(b) A – £4·75; B – £4·64; C – £5·05
Supermarket C is cheapest for basket (a).
Supermarket B is cheapest for basket (b).

E Comparing prices over a period of time

E1 (a) £7·04 (b) £20·54

E2 (a) £1·685 (b) £5·28

E3 Pupil's own local prices

3 Percentage

This chapter reviews the two basic types of percentage calculation introduced earlier in the series (calculating a percentage of a quantity, and expressing one quantity as a percentage of another) and introduces percentage increases and decreases.

A Calculating a percentage of an amount

A1 (a) 0·35 (b) 0·36 (c) 0·59
(d) 0·05 (e) 0·08

A2 (a) 0·13 (b) 0·03 (c) 0·3
(d) 0·72 (e) 0·09

A3 (a) £58·50 (b) £22·40 (c) £8·40
(d) £5·60 (e) £7·50 (f) £58·80

A4 (a) £29·70 (b) £93·50 (c) £12·60
(d) £176 (e) £16·80 (f) £481·80

A5 (a) 80 g (b) 65 g (c) 7·5 g
(d) 5 g (e) 92·5 g
Check: 80 + 65 + 7·5 + 5 + 92·5 = 250 ✓

B Percentage increases and decreases (1)

B1 £96·32

B2 Skilled workers – £9·18 per hour
Unskilled workers – £5·51 per hour

B3 (a) £83·98 (b) £523·20 (c) £32·26
(d) £67·66 ((c) and (d) to nearest penny)

B4 (a) £138 (b) £110·40 (c) £78·20
(d) £39·10

B5 (a) £72 (b) £53·71 (c) £499·20
(d) £3·80 (to nearest penny)

B6 (a) £147 (b) 17 years

C Expressing one amount as a percentage of another

C1 (a) 59% (b) 37% (c) 35% (d) 14%

C2 (a) 234 ha (b) 74 ha (c) 32% (d) 18%
(e) 50%

(f)

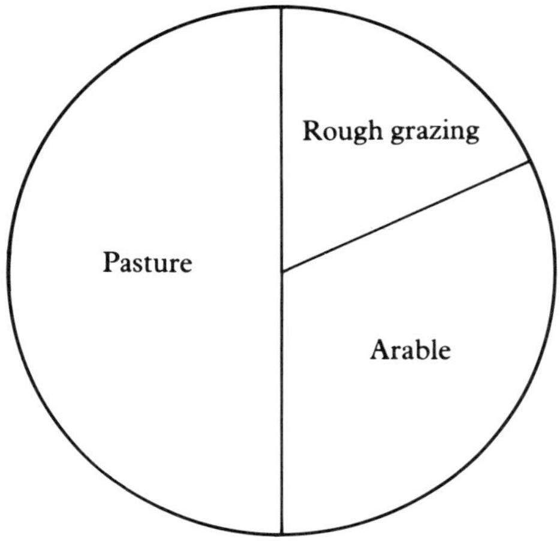

C3 (a) 341 (b) 31% (c) 21%

D Percentage increases and decreases (2)

D1 22%

D2 18%

D3 (a) £13 (b) 19%

D4 (a) 19% (b) 8% (c) 11% (d) 14%

D5 28%

D6 7%

D7 15%

D8 (a) 33% (b) 42% (c) 4% (d) 16%

D9 (a) 23% decrease (b) 29% increase
(c) 8% increase (d) 12% decrease

D10 (a) 14% (b) 17·4p per unit

E Mixed questions

E1 (a) £36·04 (b) £7·70 (c) £10·56
(d) £0·84

E2 (a) £32·24 (b) £278·80 (c) £69·55
(d) £36·19

E3 (a) 74% (b) 78% (c) 26%

E4 20%

E5 £111

E6 36%

E7 20%

E8 23%

E9 £7609

E10 13%

E11 (a) Pupil's estimate
(b) (i) 14·0 cm (ii) 8·0 cm (iii) 57%

4 Counting the cost

The material in this chapter is intended to be a stimulus to individual project work. The distinction between initial costs and running costs is often a very useful one. In the case of items with high initial costs, such as cars, motorbikes, good cameras, etc., the item retains a value in that it can be sold again. However, it may be difficult to predict what the resale will be, as it will depend very much on age, condition, and the level of demand.

5 Maps and plans

After some work on interpreting photos, plans and maps, this chapter introduces contours. This chapter will need considerable discussion.

A Plans

A1 A – Water Gate; B – Church;
C – Palace of Richard II;
D – Keep; E – Land Gate

A2 From the north-west

A3 In the afternoon

A4 (a) About 20 m (b) About 30 m

A5 About 170 m

A6 About 33 000 m^2

A7 About 8000 m^2

A8 From inside the castle

B Maps

B1 (a) A – Land Gate; B – Keep;
C – Water Gate
(b) About 250 m
(c) About 3 minutes, perhaps
(d) About 350 m

B2 (a) Sports field (b) About 3·9 km^2

B3 (a) School playing fields (b) 1 cm to 500 m
(c) About 1·5 km
(d) About 20 minutes, perhaps
(e) About 9 km^2

B4 L, M, O

B5 Portsmouth

B6 A, B, C, D, P, O

B7 L, N, B, C, D, E, G

B8 Brighton to Portchester

C Heights above sea-level

C1 The contours are more spaced out on the left-hand part, so the land must rise less quickly.

C2 (a) The path from A
(b) The path from C

C3 (a) 200 m (b) Downhill (c) Uphill
(d) From left to right

C4 (a) 300 m, 250 m, 250 m, 300 m
(b) Go downhill first, then uphill

C5 (a) Go downhill first, then uphill
(b) Go downhill all the way
(c) Go uphill all the way
(d) Go downhill first, then uphill
(e) Go downhill all the way

D Contours on OS maps

D1 90 m

D2 Uphill

D3 Driving from 510430, you go downhill for about 2 km, then rise slightly as you pass the second milestone, then downhill again.

D4 About 85 m above sea-level

D5 490408; about 70 m above sea-level

Review 1

1 Speed

1.1 6 m.p.h.

1.2 $2\frac{1}{2}$ hours

1.3 203 miles

1.4 2 hours 40 minutes

1.5 (a) 24·7 m.p.h. (to 1 d.p.)
(b) (i) 41 miles (ii) 1·25 hours
(iii) 32·8 m.p.h.
(c) 24·6 m.p.h. (to 1 d.p.)

2 Value for money

2.1 (a) 1·1 litres (b) 77·3p per litre
(c) 71·0p per litre ((b) and (c) to 1 d.p.)
(d) The pack of 4 cans is better value for money.

3 Percentage

3.1 (a) £75 (b) £68 (c) £28·80 (d) £11·70

3.2 £5773 (to nearest £)

3.3 (a) 72% (b) 42% (c) 42%

3.4 38% (to nearest 1%)

3.5 26%

3.6 13%

5 Maps and plans

5.1 (a) (i) C, F (ii) D, B, J (iii) D, E
(b) Dartford to Maidstone, Maidstone to Tonbridge, Tonbridge to Maidstone

5.2 The south side is the steepest.

6 Rates

This chapter is concerned with calculations involving various kinds of rate: rates of pay per hour, typing speeds, etc.

A Rates

A1 18 000 cubic feet

A2 315 miles

A3 £94·25

A4 (a) £207·20 (b) 8 hours

A5 45 words per minute

A6 36 miles per gallon

A7 (a) 67 litres per minute
(b) 12 litres per minute

A8 0·4 litre per minute

A9 £2·75 per hour

A10 76 beats per minute

A11 84 beats per minute

B Calculations involving rates

B1 8 litres

B2 (a) 140 copies per minute
(b) 71·4 minutes (to 1 d.p.)

B3 The turntable is too slow – 31·6 revolutions per minute.

B4 $3\frac{1}{2}$ hours

B5 No. She has enough fuel for 368 miles more, assuming the consumption rate remains the same. Pupils will suggest other equivalent explanations.

B6 2 hours 6 minutes, or 2·1 hours

Money matters: VAT

The rate of VAT may change, as also the range of goods and services to which it applies, so up-to-date information is necessary. For a time there were different rates for different kinds of goods, and this practice could be revived. Small businesses whose yearly turnover is below a certain amount do not have to charge VAT; the threshold may be altered from time to time.

1 At 15% VAT
(a) £11·50 (b) £23 (c) £57·50 (d) £63·25 (e) £67·28

7 Mixing and sharing

This chapter deals with mixing and sharing quantities in a given ratio, expressed in the form '2 to 3', for example.

A Mixing in a given ratio

A1 (a) 6 litres (b) 15 litres (c) 10·5 litres

A2 (a) 4 litres (b) 6 litres (c) 0·5 litre

A3 (a) 6 litres (b) 12 litres (c) 15 litres

A4 (a) 9 litres (b) 16 litres

A5 (a) 7·5 litres (b) 3 litres

A6 (a) 6 litres (b) 15 litres

A7 (a) 9 litres (b) 16 litres

A8 5 to 3, 2 to 1, etc.

A9 3 litres of red and 9 litres of white

A10 (a) 4 red, 6 white (b) 10 red, 15 white
(c) 1 red, 1·5 white (d) 8 red, 12 white
(e) 12 red, 18 white
(f) 0·4 red, 0·6 white

A11 (a) 12 red, 9 white (b) 2 red, 1·5 white
(c) 20 red, 15 white

B Sharing in a given ratio

B1 Sadia £120, Tina £180

B2 Sadia £168, Tina £252

B3 Sadia £160, Tina £320

B4 (a) Sadia £61, Tina £122
(b) Sadia £172·33, Tina £344·67
(c) Sadia £69·67, Tina £139·33
(d) Sadia £90·83, Tina £181·67

B5 (a) Sadia £108, Tina £144
(b) Sadia £62·57, Tina £83·43
(c) Sadia £135·54, Tina £180·71
(d) Sadia £74·31, Tina £99·07
(e) Sadia £92·61, Tina £123·48

B6 (a) In the ratio 4 to 5
(b) Sadia £156·18, Tina £195·22

C Sharing between three

C1 Share in the ratio 3 to 4 to 3.
Kate £135, Cheryl £180, Pauline £135

C2 Manoj £75, David £100, Ronnie £125

C3 7·5 litres blue, 10 litres yellow, 2·5 litres white

D Miscellaneous problems on mixing and sharing

D1 (a) 4750 ml = 4·75 litres
(b) 5 litres PVA, 25 litres water
(c) 7·5 litres PVA, 7·5 litres water
(d) 2·5 litres PVA, 0·5 litre water

D2 (a) Ruth £360, David £240
(b) Ruth £342·86, David £257·14 (to nearest p)
(c) Ruth £320, David £280
(d) Ruth £312, David £288

D3 Karen £80, Judith £100

Money matters: insurance

Vehicle insurance and holiday insurance are likely to be of some interest to pupils. As in the case of the other financial topics, there is no adequate substitute for collecting real data. It is instructive to compare the premiums charged and the coverage offered by different companies for similar classes of insurance, for example a two-week stay in Europe.

1 People on winter sports holidays have more accidents.

2 (a) £135 (b) £126

8 Graphs and charts

After some work on getting information from bar charts, this chapter goes on to examine ways in which graphs and charts may mislead.

A Bar charts

A1 (a) About 148 000 (b) About 102 000
(c) About 17 000

A2 About 30%

A3 (a) About 10% (b) About 60% or 70%

A4 About 60%

A5 60%

A6 (a) Greater London (b) Tyne and Wear

A7 2992 people per km^2
So there is about $1\frac{1}{2}$ times as many people per km^2 in the West Midlands as there is in Greater Manchester.

A8 (a)

Region	**Population density in people per km^2**
Greater London	4307
West Midlands	2992
Merseyside	2316
Tyne and Wear	2093
Greater Manchester	2012
West Yorkshire	991
South Yorkshire	833

All densities are given to the nearest whole number.

(b)

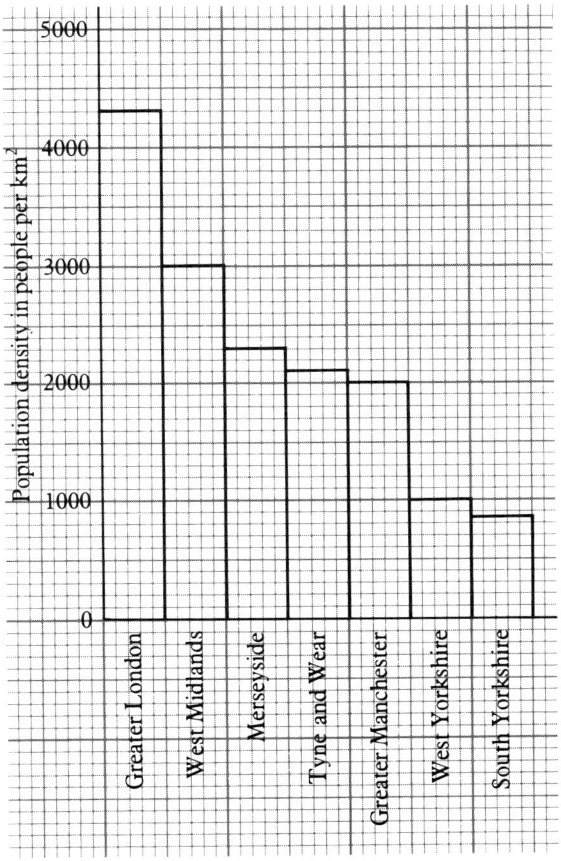

A9 (a) Bedfordshire (b) Northumberland

(c)

County	Population density in people per km²
Bedfordshire	403
Oxfordshire	213
Cambridgeshire	175
Cornwall	124
Shropshire	109
Lincolnshire	93
Cumbria	69
Northumberland	59

All densities are given to the nearest whole number.

B Misleading charts

This section will need considerable discussion.

B1 The 1987 bar is four times the area of the 1986 bar, but the profit is only twice as much.

B2 The larger loaf in chart B is at least four times the area of the smaller loaf, so this is an unfair picture. Chart A gives a fairer picture by using a standard symbol to represent 100 000 loaves.

B3 The charts do not give a fair picture. The symbols give the impression of an increase of about four times, not twice.

B4 Plantin Software is growing faster, despite the contrary first impression given by the graphs.

B5 The vertical axis does not stand at zero. The scale has been chosen to make it look as though sales have nearly doubled each year, but in fact they have increased by only about 4% over the four years.

B6 Pupil's own misleading chart

Money matters: foreign exchange

As is the case of the other topics in this series, up-to-date information is needed to bring the topic to life. Although most people are directly interested in exchange rates only when they are going abroad, there remain in the background wider issues, such as what determines exchange rates and what causes them to change.

1 (a) 593F (b) 948·8F

2 £42·16 (to nearest p)

3 £2·95, £4·55 and £6·07 (all to nearest p)

Review 2

6 Rates

6.1 (a) 960 litres per day
(b) After 13 more days just 120 litres of water are left.

6.2 (a) 272 miles (b) 6·25 gallons

6.3 (a) 63 hours (b) 3·78 units
(c) 18·9p, or 19p to the nearest penny

7 Mixing and sharing

7.1 (a) Rosie £96, Chris £64
(b) £51 (c) £56

7.2 (a) 5 to 3 (b) £3750 and £2250

8 Graphs and charts

8.1 British Isles about 60%,
Far East about 20%,
Central Africa about 10%

8.2 The area of the larger balloon is about four times the area of the smaller balloon, so this would suggest an increase of about 300%, not 25%.

8.3 The chart gives the impression that the number of CEDRIC computers sold in 1987 is about twice the number sold in 1986. In fact the increase is only about 20%.

9 Squares and square roots

This short chapter introduces the idea of square root, and shows how the 'squares graph' (the graph of $y = x^2$, not described as such in the text) may be used to estimate non-integral square roots. It may be possible to introduce the decimal search method of calculating a square root at this point. (Of course, there are likely to be some pupils in the class whose calculators have a square root key.)

A Square roots

A1 (a) 36 (b) 49 (c) 64 (d) 81 (e) 100

A2 (a) 529 (b) 3249 (c) 164 025
(d) 698 896

A3 (a) 2209 (b) 105 625 (c) 250 000

A4 (a) 6 (b) 2 (c) 9

A5 (a) 8 × 8 is (64), so 8 is the square root of (64).
(b) The square of 10 is (100), so 10 is the square root of (100).
(c) 7 × 7 is (49), so (49) is the square of (7), and (7) is the square root of (49).

B The squares graph

B1 (a) 0·2 (b) 2

B2 (a) About 6·3 (b) $6{\cdot}3^2 = 39{\cdot}69$

B3 About 5·5. $5{\cdot}5^2 = 30{\cdot}25$

B4 (a)

Number	0 ... 8 9 10
Square of number	0 ... 64 81 100

The graph will, of course, be an extended version of that on page 71.

(b) About 8·7 (c) $8{\cdot}7^2 = 75{\cdot}69$

Money matters: saving

Real information about savings schemes can be collected and schemes compared. In deciding on a scheme the advantage of a higher rate of interest has to be weighed against the disadvantages of some of the conditions relating to the scheme (for example, the saver may need to invest a minimum sum, or may need to give a month's notice of withdrawal).

1

	£
1st June 87	500·00
Interest 87–88	+ 40·00
1st June 88	540·00
Interest 88–89	+ 43·20
1st June 89	583·20
Interest 89–90	+ 46·66
1st June 90	629·86
Interest 90–91	+ 50·39
1st June 91	680·25

2

	£
1st Sept 87	200·00
Interest 87–88	+ 24·00
1st Sept 88	224·00
Interest 88–89	+ 26·88
1st Sept 89	250·88
Interest 89–90	+ 30·11
1st Sept 90	280·99

3 In five years' time the bike is likely to cost more than £750.

10 Arranging and selecting

Section A looks at some problems of timetabling activities. The other sections are concerned with listing different possible arrangements of items. Section D introduces branching diagrams (tree diagrams) as an aid to methodical counting.

A Timetables

A1 Here is one possible solution.

	Monday	**Tuesday**	**Wednesday**	**Thursday**	**Friday**
Group A	Map reading	Water safety	Canoeing	Hill walking	Athletics
Group B	Water safety	Canoeing	Athletics	Map reading	Hill walking
Group C	Athletics	Map reading	Hill walking	Water safety	Canoeing

A2 Here is one possible solution.

Day	**1**	**2**	**3**	**4**	**5**
Rachel	✓	✓			✓
Sadie	✓	✓	✓		
Kevin	✓	✓		✓	
Martin			✓	✓	✓
Peter			✓	✓	✓

A3 (a) No, it cannot be done.
(b) Yes, it can now be done, for example,

Week	**1**	**2**	**3**	**4**	**5**	**6**
Janet	✓	✓		✓		
Rajesh	✓	✓			✓	
Patrick			✓	✓		✓
Gillian			✓		✓	✓

A4 Here is one possible solution.

1st game	A plays B.	C, D and E rest.
2nd game	C plays D.	A, B and E rest.
3rd game	A plays E.	B, C and D rest.
4th game	B plays C.	A, D and E rest.
5th game	D plays E.	A, B and C rest.
6th game	A plays C.	B, D and E rest.
7th game	B plays D.	A, C and E rest.
8th game	C plays E.	A, B and D rest.
9th game	A plays D.	B, C and E rest.
10th game	B plays E.	A, C and D rest.

B Choosing pairs

B1

Day	1	2	3	4	5	6	7	8
Main course	S	S	H	H	F	F	P	P
Pudding	R	I	R	I	R	I	R	I

On 8 days

B2 (a) There are many possible solutions.
(b) For 9 days

B3 (a) There are many possible solutions.
(b) No, there are 12 different outfits.

C Picture cards

C1 PRQS, PRSQ, PSQR, PSRQ

C2 QPRS, QPSR, QRPS, QRSP, QSPR, QSRP
giving a total of 6 different pictures with Q on the left

C3 (a) 6 different pictures
(b) 24 different pictures

C4 A2Y, B2Z, B3Y

D Listing possibilities

D1 (a)

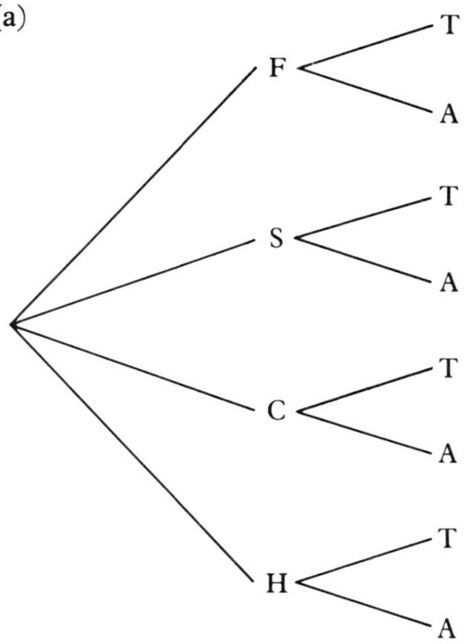

(b) FT, FA, ST, SA, CT, CA, HT, HA

D2 (a)

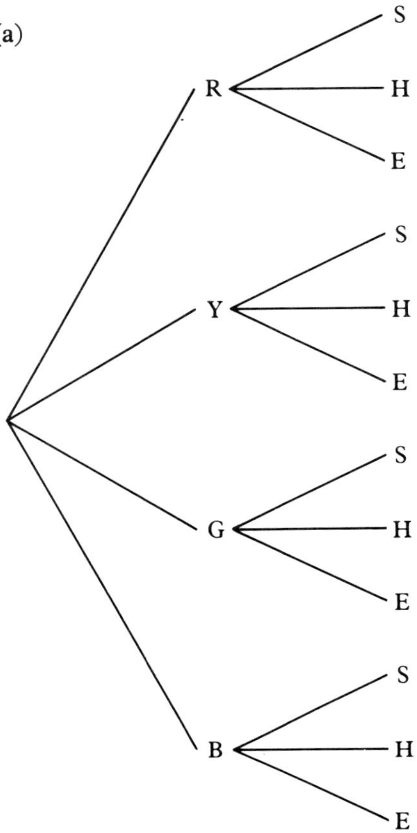

(b) There are 12 different combinations:
RS, RH, RE, YS, YH, YE, GS, GH, GE, BS, BH, BE

D3 (a)

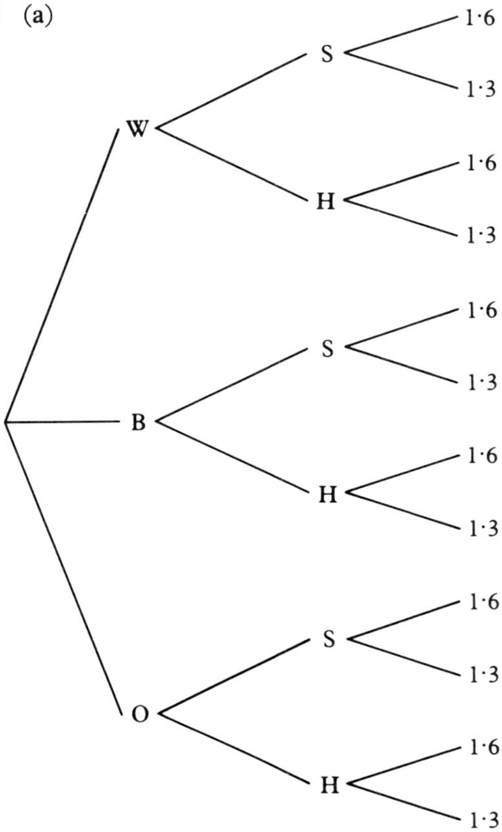

(b) There are 12 combinations.

11 Fractions

Fractions with denominators 2, 4, 8, 16, ... are still used sometimes in measurement. This chapter deals with equivalence, addition and subtraction of halves, quarters, eighths and sixteenths, and finishes with a brief section on equivalence of fractions in general.

A Halves, quarters, eighths, ...

A1 (a) $\frac{3}{8}$ (b) $\frac{3}{4}$ (c) $\frac{9}{16}$
(d) $\frac{1}{4}$ (e) $\frac{7}{8}$ (f) $\frac{5}{16}$

A2 (a) $\frac{1}{2}$ (b) $\frac{1}{4}$ (c) $\frac{3}{8}$ (d) $\frac{3}{4}$ (e) $\frac{1}{4}$

A3 (a) $\frac{12}{16}$ (b) $\frac{10}{16}$ (c) $\frac{8}{16}$ (d) $\frac{14}{16}$

A4 (a) $\frac{4}{16}''$ (b) $\frac{6}{16}''$ (c) $\frac{5}{16}''$

A5 Rods B, D and E are too big.

A6 (a) $\frac{5}{16}$ (b) $\frac{13}{16}$ (c) $\frac{13}{16}$ (d) $\frac{5}{8}$
(e) $\frac{13}{16}$ (f) $\frac{7}{16}$ (g) $\frac{1}{8}$

A7 (a) 19 mm (b) 22 mm (c) 17 mm
(d) 5 mm (e) 24 mm

B Writing a fraction in its simplest form

B1 (a) $\frac{2}{3}$ (b) $\frac{3}{5}$ (c) $\frac{4}{5}$ (d) $\frac{3}{4}$
(e) $\frac{2}{5}$ (f) $\frac{2}{3}$

B2 $\frac{20}{32} = \frac{5}{8}$

Money matters: borrowing

Mortgages are of course a common instance of loans repaid in instalments, but are likely to be less real to pupils than hire purchase.

1

	£
1st Oct 89	135·50
Interest 89–90	+ 20·83
	155·83
Repayment	− 60·00
1st Oct 90	95·83
Interest 90–91	+ 14·37
	110·20
Repayment	− 60·00
1st Oct 91	50·20
Interest 91–92	+ 7·53
	57·73
Repayment	− 57·73
	00·00

12 Metric units

This chapter reviews the metric units of length, volume (capacity) and weight, and conversions between units (e.g. centimetres to metres).

A Multiplication and division by 10, 100, 1000: a review

A1 (a) 68 (b) 7·2 (c) 8·59 (d) 127·4 (e) 0·81

A2 (a) 346 (b) 2325·1 (c) 870 (d) 1608 (e) 4370

A3 (a) 1408 (b) 12 632 (c) 837·5 (d) 8005 (e) 62

A4 (a) 68·4 (b) 73·5 (c) 5·4 (d) 200·7 (e) 420 (f) 0·732 (g) 6090 (h) 2320 (i) 83

A5 (a) 5·7 (b) 0·57 (c) 0·628 (d) 1·15 (e) 0·906

A6 (a) 3·84 (b) 16·35 (c) 0·696 (d) 0·085 (c) 3·058

A7 (a) 6·842 (b) 0·375 (c) 0·0163 (d) 0·1481 (e) 0·0072

A8 (a) 6·34 (b) 0·634 (c) 0·0634 (d) 0·807 (e) 0·26 (f) 0·483 (g) 7·62 (h) 9·84 (i) 0·0372

A9 (a) 86 (b) 0·732 (c) 480 (d) 6·825 (e) 0·68 (f) 0·3124 (g) 1·835 (h) 0·605 (i) 710

B Units of length

B1 (a) 160 mm (b) 8 mm (c) 53 mm (d) 18·5 mm

B2 (a) 2·6 cm (b) 0·5 cm (c) 31·5 cm (d) 7·2 cm

B3 (a) 160 cm (b) 167 cm (c) 2582 cm (d) 9 cm

B4 (a) 1·83 m (b) 0·56 m (c) 0·07 m (d) 13·2 m

B5 (a) 350 cm (b) 8 mm (c) 32·54 m (d) 3·2 cm (e) 86 cm (f) 188 mm

B6 (a) 5000 m (b) 5800 m (c) 630 m (d) 1862 m

B7 (a) 4·35 km (b) 0·232 km (c) 0·05 km (d) 0·75 km

B8 (a) 632 mm (b) 0·25 km (c) 13 500 m (d) 8 mm (e) 3·84 m (f) 2·7 cm (g) 0·065 km (h) 0·82 m (i) 0·03 km

B9 1·6 km

B10 700 m

B11 (a) 1300 m (b) 260 pipes

B12 9 lengths

C Units of volume and weight

C1 (a) 3000 ml (b) 3200ml (c) 800 ml (d) 882 ml

C2 (a) 0·6 litre (b) 0·842 litre (c) 0·073 litre (d) 0·009 litre

C3 (a) 0·45 litre (b) 2300 ml (c) 60 ml (d) 0·048 litre

C4 (a) 1·356 litres (b) 880 cm^3 (c) 0·42 litre (d) 51 cm^3

C5 (a) 8000 g (b) 650 g (c) 4200 g (d) 900 g

C6 (a) 0·382 kg (b) 0·056 kg (c) 0·08 kg (d) 1·35 kg

C7 (a) 2500 g (b) 0·75 kg (c) 660 g (d) 1502 g (e) 1·8 kg (f) 0·02 kg

C8 (a) 0·4 litre (b) 6200 g (c) 8200 ml (d) 0·18 kg (e) 70 g (f) 1050 ml (g) 3·75 litres (h) 0·125 kg

C9 (a) (i) 500 g (ii) 250 g (iii) 750 g (b) (i) 500 ml (ii) 250 ml (iii) 750 ml (c) 14 packs

D Mixed questions

D1 40 bottles

D2 50 portions

D3 125 lengths

D4 15 mm

D5 (a) 50 coins (b) 50 000 coins (c) £500

D6 (a) 15 cm (b) 3 m

D7 86·4 m

13 Formulas and graphs

After a brief review of algebraic notation, this chapter gives examples of graphing formulas, each of which refers to some concrete relationship.

A The language of algebra: a review

A1 (a) 20 (b) 6 (c) 6 (d) 30 (e) 80 (f) 120

A2 (a) 23 (b) 26 (c) 22 (d) 7 (e) 13 (f) 18 (g) 5 (h) 27

A3 (a) 11 (b) 7 (c) 50 (d) 4 (e) 300

A4 (a) 50 (b) 12 (c) 1 (d) 39 (e) 21 (f) 20 (g) 50 (h) 90 (i) 23 (j) 2

A5 (a) 4 (b) 4 (c) 8 (d) 2 (e) 14 (f) 10 (g) 12

A6 (a) 16 (b) 6 (c) 30 (d) 9 (e) 4

A7 (a) 10 (b) 30 (c) 36 (d) 18 (e) 9 (f) 1

B Substituting numbers into formulas

B1 (a) 5 (b) 6 (c) 3·6 (d) 2·94

B2 (a) 3·87 (b) 10·75 (c) 18·1675

B3 (a) 16·5 (b) 136 (c) 110 (d) 27·5

B4 (a) 1131 (b) (i) 1734 (ii) 820
(to nearest whole number)

C Graphs from formulas

C1 (a)

S	0	10	20	30	40	50	60	70
B	0	1·7	6·7	15	26·7	41·7	60	81·7

(b)

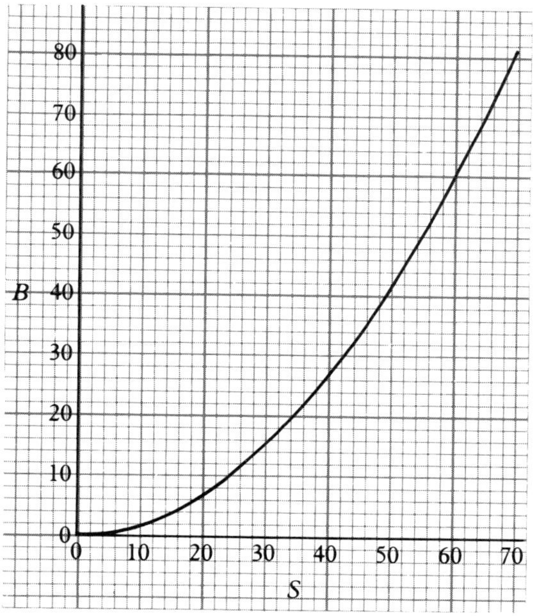

(c) About 20 or 21
(d) About 55 m.p.h.

C2 (a) 5

(b)

S	0	10	20	30	40	50	60	70
D	0	5	13·3	25	40	58·3	80	105

(c)

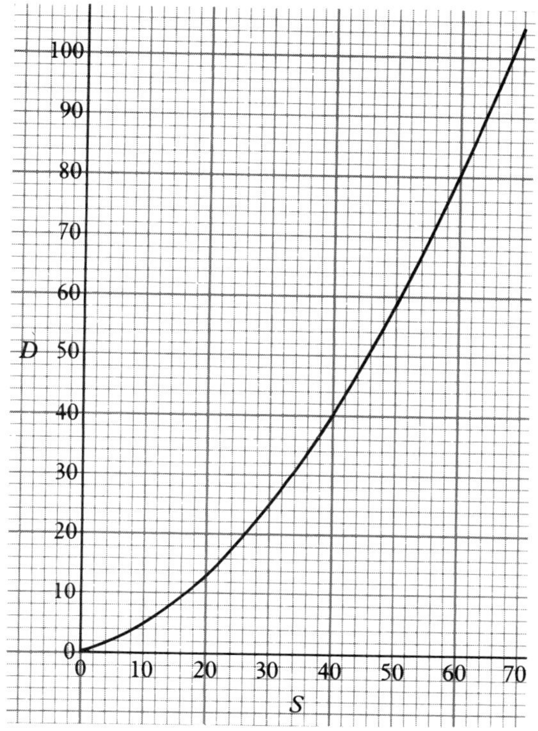

(d) About 46 m.p.h.

C3 (a) 35 (b)

t	0	1	2	3	4	5
h	80	75	60	35	0	$^{-}45$

(c)

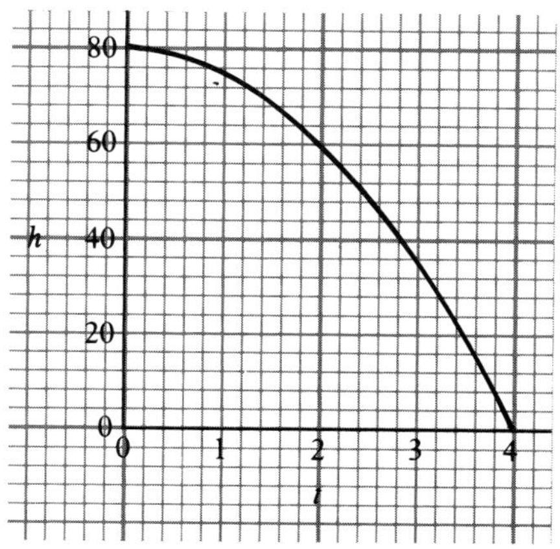

(d) The stone speeds up.
(e) About 2·8 seconds

C4 (a)

c	0	10	20	30	40
f	32	50	68	86	104

(b)

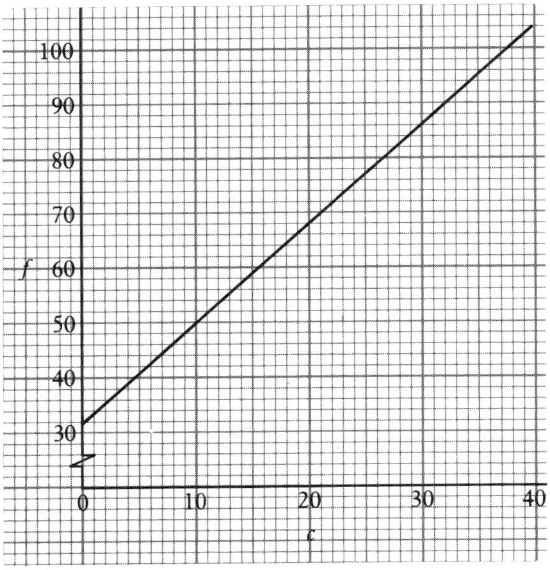

(c) About 16 °C

C5 (a)

S	20	30	40	50	60
T	6	4	3	2·4	2

(b)

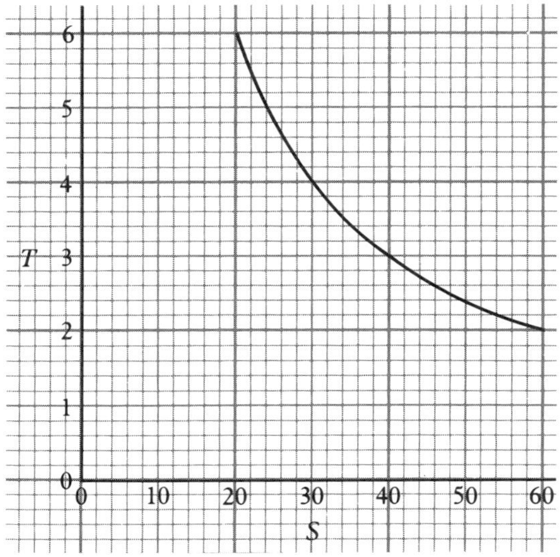

(c) T gets smaller and smaller.
(d) T gets larger and larger.
(e) Cannot divide by 0.

Review 3

9 Squares and square roots

9.1 (a) 25 (b) 49 (c) 100

9.2 (a) 4 (b) 1 (c) 8

9.3 (a) ③ is the square root of 9.
(b) ㊇1 is the square of 9.
(c) 4 is the square of ②.
(d) 7 is the square root of ㊾ .

10 Arranging and selecting

10.1 27 different numbers

10.2 15 combinations

11 Fractions

11.1 (a) $\frac{6}{16}$ (b) $\frac{12}{16}$ (c) $\frac{8}{16}$ (d) $\frac{14}{16}$

11.2 (a) $\frac{11}{16}$ (b) $\frac{7}{8}$ (c) $\frac{7}{16}$ (d) $\frac{3}{16}$

11.3 (a) $\frac{3}{4}$ (b) $\frac{4}{5}$ (c) $\frac{2}{3}$ (d) $\frac{9}{10}$ (e) $\frac{7}{8}$

12 Metric units

12.1 (a) 5·6 cm (b) 0·35 m (c) 3600 m
(d) 0·45 km (e) 0·95 kg (f) 70 ml

12.2 15 portions

12.3 200 pieces

13 Formulas and graphs

13.1 (a) 17 (b) 25 (c) 34 (d) 50 (e) 12
(f) 0 (g) 4 (h) 2 (i) 6·5 (j) 6

13.2 (a)

t	0	1	2	3	4	5
d	0	5	20	45	80	125

(b)

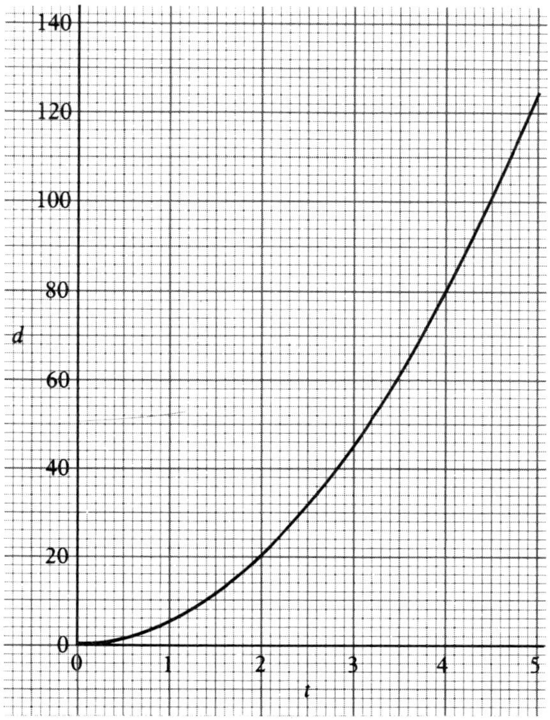

(c) (i) About 34 m (ii) About 3·2 seconds

General review

Sections 1 to 23 give an opportunity for topic-by-topic revision, which may reveal areas of weakness where further attention is necessary. Section M is made up of questions taken form the pilot 16+ examination based on the SMP 11–16 course. Questions M1 to M33 appeared on paper 1 and M34 to M52 on paper 2 (although not all on the same occasion in either case).

1 Whole numbers and decimals

1.1 576 bottles

1.2 276 tins

1.3 765 rabbits

1.4 300 tiles

1.5 (a) 36 boxes (b) 85 jars

1.6 (a) Yes (b) No (c) No

1.7 (a) £5·44 (b) £1·53 (c) £42·16

1.8 (a) £734 000 (b) £218 000 (c) £28 000

1.9 (a) 0·833 kg (to 3 d.p.)
(b) 1·2 kg

1.10 Standard: 6·32 ml for 1p or 0·158p per ml
Large: 6·00 ml for 1p or 0·167p per ml
So the standard cup is better value.

1.11 832 labels

1.12 40 doses

1.13 4·6 cm

1.14 6·3 6·08 4·289 1·7 0·367 0·0295 0·009 88

1.15 (a) 26000 (b) 35·79 (c) 0·05 (d) 7·44

1.16 The population of Bristol is about ③ times that of Reading.

1.17 184

1.18 £83·37 (to the nearest p)

2 Percentage

2.1 (a) '... I'll reduce the price by 25%.'
(b) '70% of students finish the course.'
(c) '40% of government expenditure goes on education.'

2.2 (a) £36 (b) £1·96 (c) £71·50

2.3 65%

2.4 (a) 43% (b) 58% (c) 7% (d) 4%

2.5 (a) 75% (b) 92% (to nearest 1%)
(c) 6 carat gold

2.6 £59·89

2.7 (a) £97·75 (b) £384·25

2.8 (a) £35 (b) £12·95 (c) £4·89
(d) £17·49

2.9 (a) £125 (b) 7%

2.10 (a) 19%
(b) Carol's increase was by 24%, which is larger.

3 Unitary method and ratios

3.1 (a) £9·75 (b) £136·50

3.2 302·5 cubic metres

3.3 170 g

3.4 Sheila £257·14, Ramesh £192·86

3.5 (a) 7·5 kg (b) £1·90

4 Fractions

4.1 (a) £63 (b) £44·80 (c) £144 (d) £2·94

4.2 62·5 miles

4.3 0·3125

4.4 $\frac{1}{2}''$, $\frac{5}{8}''$, $\frac{11}{16}''$, $\frac{3}{4}''$, $\frac{15}{16}''$

4.5 $\frac{13}{16}''$

4.6 (a) 6·4 mm (b) 19·1 mm (c) 15·9 mm
(d) 20·6 mm

5 Time

5.1 (a) 1715 (b) 2245 (c) 2005

5.2 (a) 3:10 p.m. (b) 9:53 p.m. (c) 6:20 a.m.

5.3 (a) 36 minutes (b) 1210 (c) 1957

6 Rates

6.1 (a) 2750 miles (b) 9 hours

6.2 364 m.p.h.

6.3 $3\frac{3}{4}$ hours

6.4 69 words per minute

6.5 (a) 65p per page (b) £28·60

6.6 £1·75

6.7 97000 frames

7 Negative numbers

7.1 (a) 8 degree fall (b) 11 degree fall
(c) 9 degree rise (d) 23 degree rise

7.2 (a) $^{-}7$°C (b) 2°C

8 Formulas

8.1 (a) 75 (b) 23·52

8.2 (a) 50 (b) 20 (c) 40
(d) 26 (e) 30 (f) 2·5

8.3 (a) 6 mg (b) 24 mg

8.4 3072 kg

9 Graphs

9.1

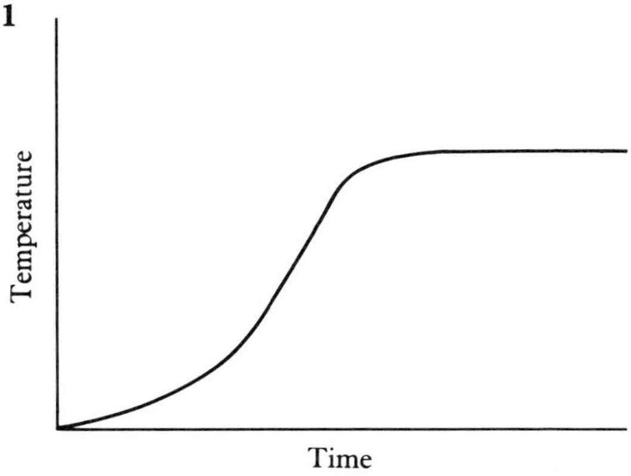

9.2 (a) 50 minutes (b) About 12 miles
(c) The coach was stationary, possibly a rest stop.
(d) The coach slowed down as it neared Birmingham, possibly as a result of leaving the motorway.

9.3 (a)

u	0	10	20	30	40	50
h	0	10	40	90	160	250

(b)

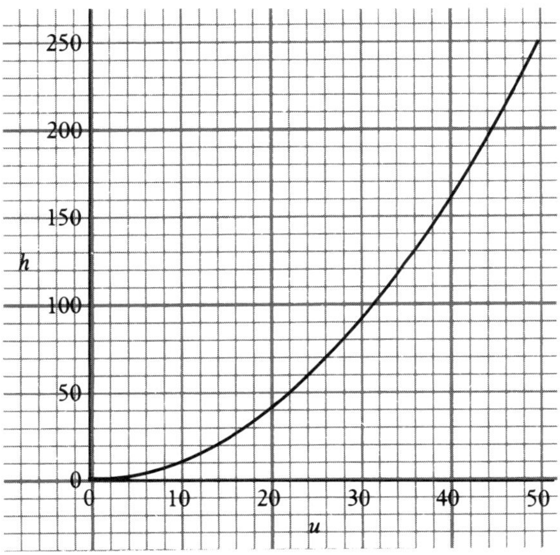

(c) u is about 35 when h is 120.

9.4 (a)

w	0	1000	2000	3000	4000	5000
h	180	160	140	120	100	80

(b)

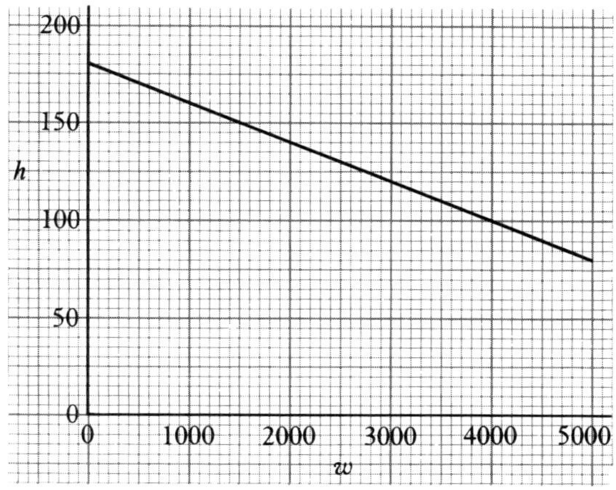

(c) 3500 kg

10 Coordinates

10.1 (a)

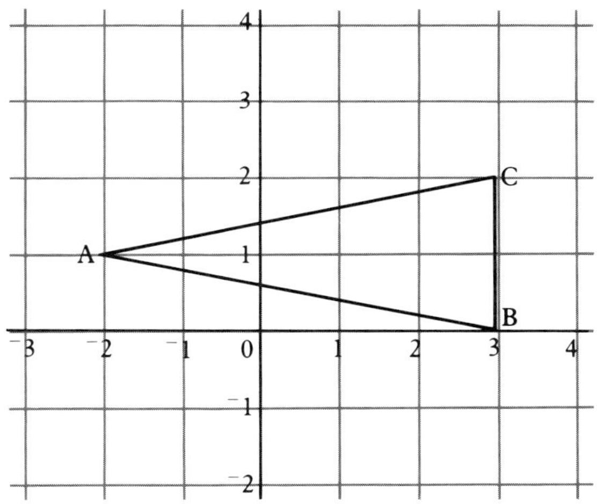

(b) Isosceles triangle

10.2

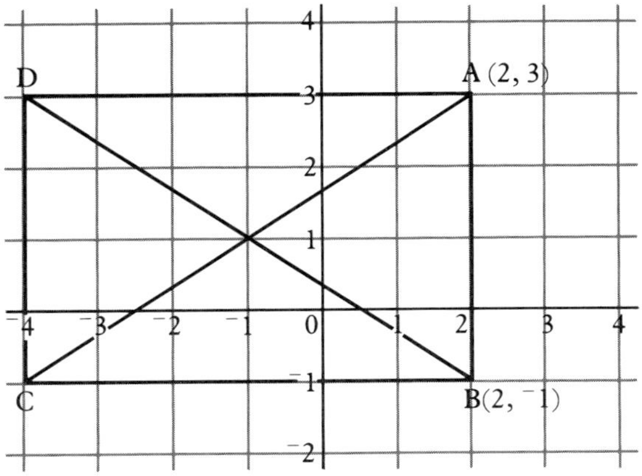

(a) $(^{-}4, ^{-}1)$ (c) $(^{-}4, 3)$ (d) $(^{-}1, 1)$

10.3

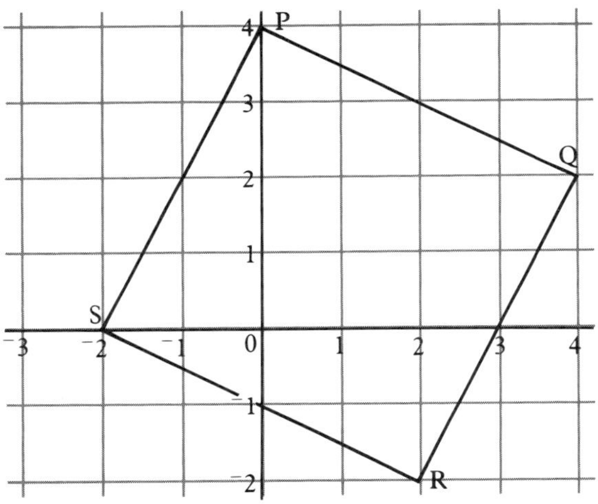

(a) P (0, 4), Q (4, 2), R $(2, ^{-}2)$
(c) S $(^{-}2, 0)$ (d) (1, 1) (e) 45°

10.4 (a)

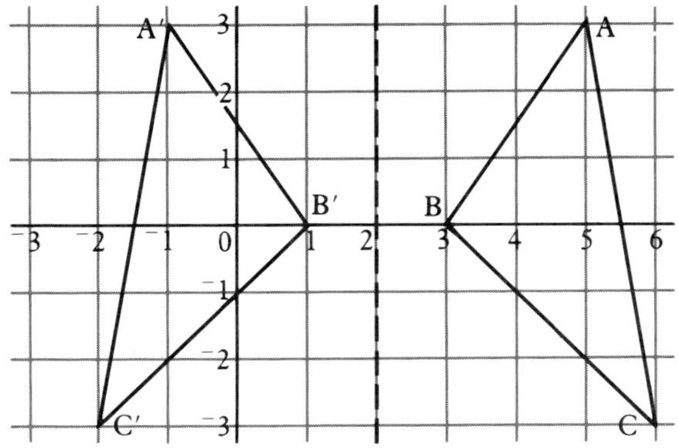

(b) A′ $(^{-}1, 3)$, B′ (1, 0), C′ $(^{-}2, ^{-}3)$

11 Area and perimeter

11.1 (a) 360 m^2 (b) 625 m^2

11.2 (a) 400 km^2
(b) (i) About 20 000 km^2
(ii) About 15 000 km^2
(c) (i) About 140 people per km^2
(ii) About 320 people per km^2

11.3 (a) 48 m (b) 80 m (c) 32 m

11.4 (a) 56 cm^2 (b) 13·5 cm^2 (c) 11·4 cm^2

11.5 (a) 42·8 m (b) 102·01 m^2 (c) 102 m^2
(d) About 23 litres

12 The circle

12.1 (a) 7 panels (b) 9 panels

12.2 23·6 cm

12.3 (a) 40·2 cm (b) 24·4 cm (c) 55·3 cm

13 Angles

13.1 $a = 32°$, $b = 130°$, $c = 37°$, $d = 161°$

13.2 (a) $a = 29°$ (b) $b = 147°$
(c) $c = 109°$, $d = 71°$, $e = 109°$

13.3 (a) E, F (b) L, P (c) H, J (d) D, B

14 Scale drawing

14.1 (a)

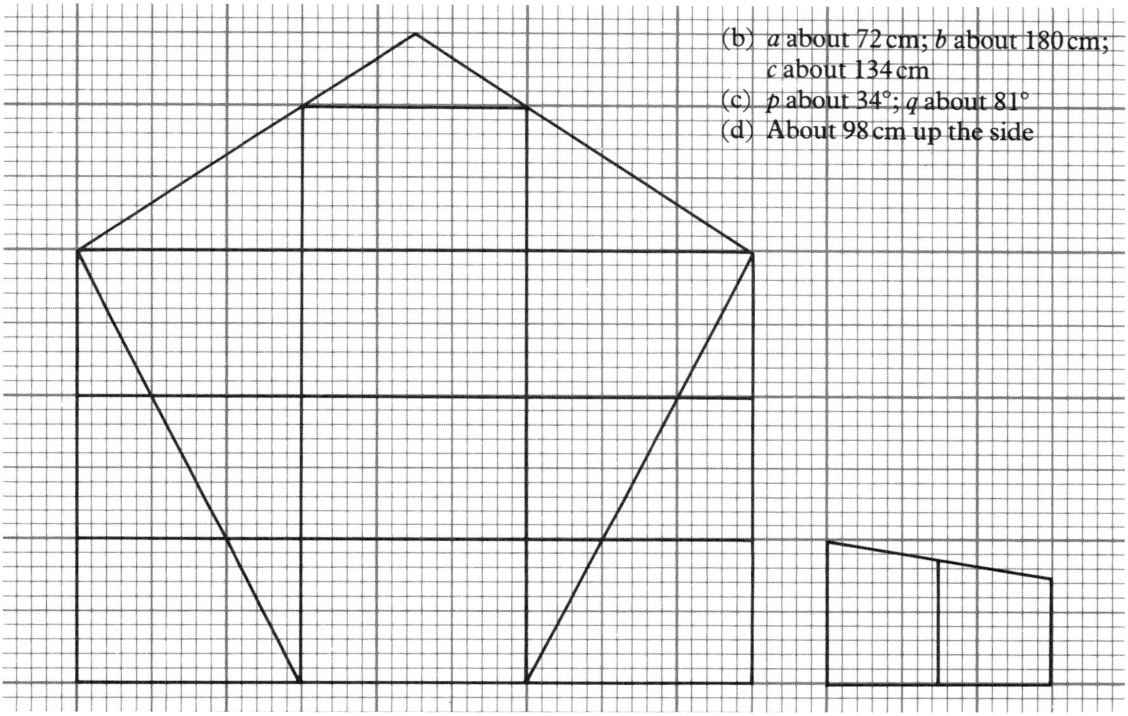

(b) a about 72 cm; b about 180 cm; c about 134 cm
(c) p about 34°; q about 81°
(d) About 98 cm up the side

14.2

Scale 1 cm to 10 m

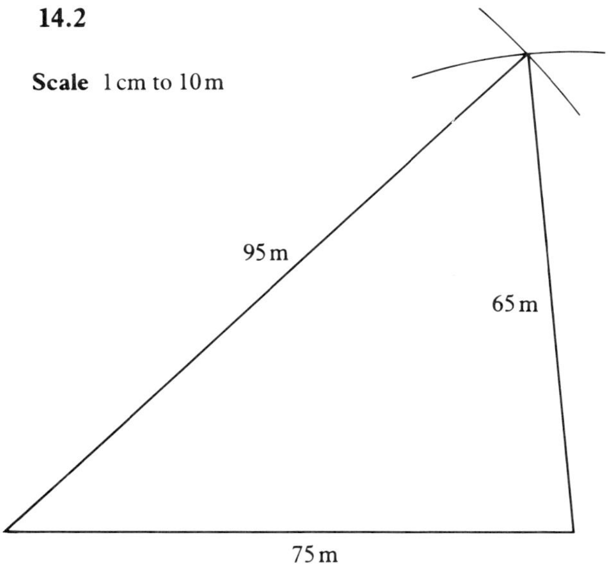

(a) Pupil should describe the use of compasses to draw triangles with three known sides.
(b) No, it is not possible.

15 Maps

15.1 (a) 1704
(b) (i) 681039 (ii) 694036 (iii) 718030
(iv) 673052 (v) 696028 (vi) 723051

15.2 (a) Between 100 m and 200 m
(b) Between 300 m and 400 m
(c) Between 100 m and 200 m
(d) Between 0 m and 100 m
(e) Between 100 m and 200 m

15.3 The path from Inlet House is steeper. The contours are closer together.

15.4 5·5 km

15.5 About 15 or 16 km

16 Symmetry

16.1

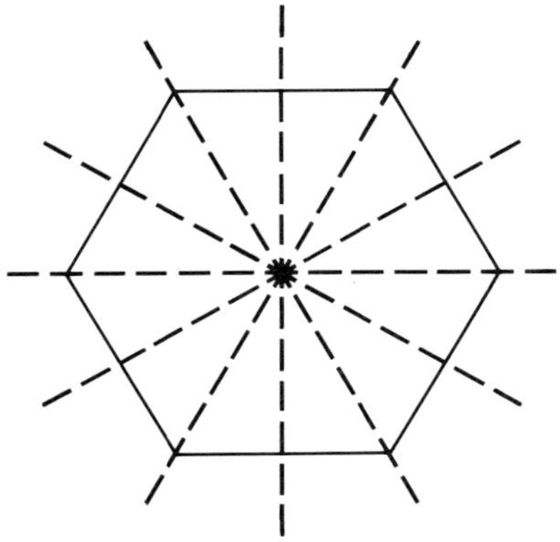

16.2 (a) Pupil's drawing of isosceles triangle
(b) Pupil's drawing of kite or arrowhead

16.3

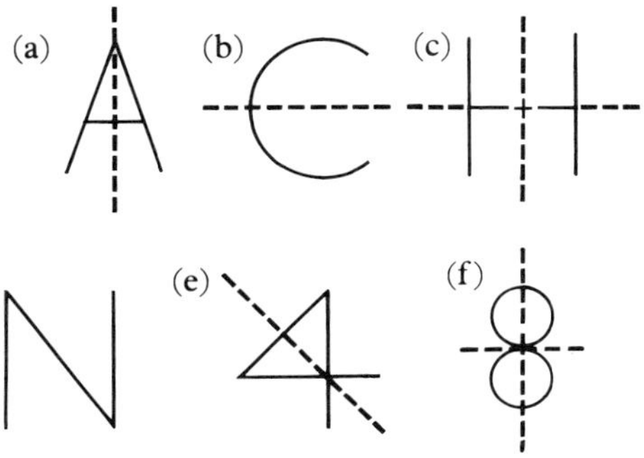

16.4 C

16.5

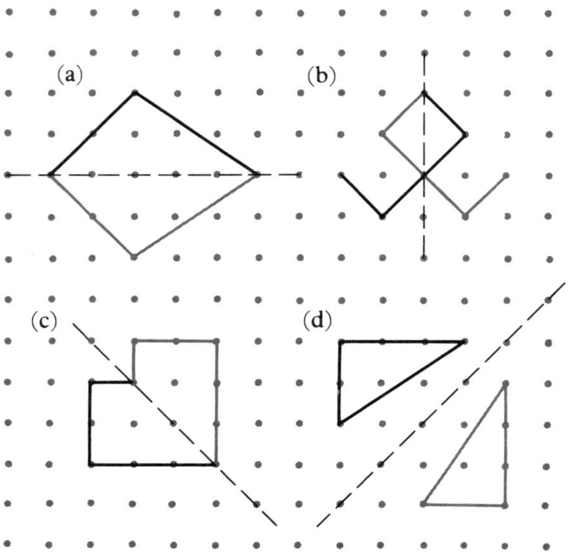

17 Three dimensions

17.1

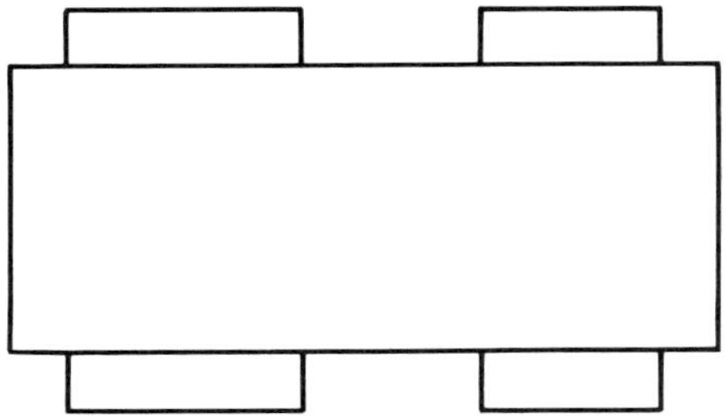

17.2 (a) South-east (b) West (c) North-east
(d) North-west (e) East

17.3 (a) Here is one possible net:

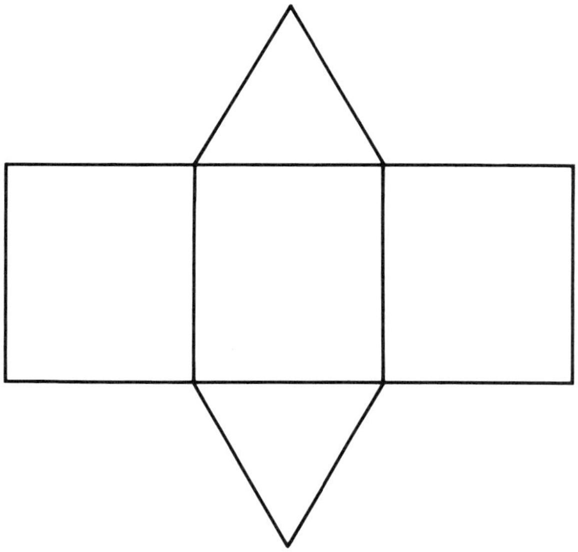

(b)

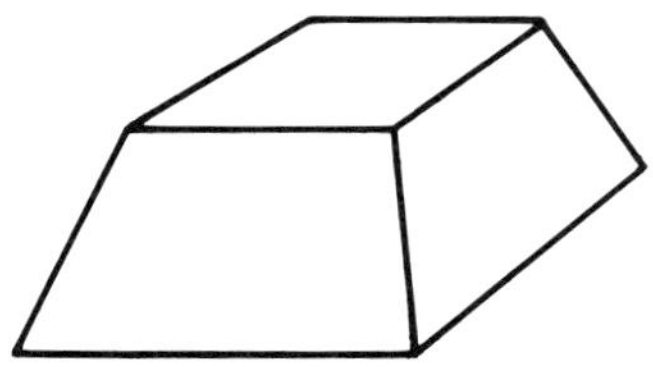

18 Metric units

18.1 (a) 3·42 km (b) 65·2 cm (c) 0·759 km
(d) 10·6 m

18.2 (a) 0·35 kg (b) 600 g (c) 1·08 kg
(d) 4200 g

18.3 (a) 0·362 litre (b) 650 ml (c) 0·025 litre
(d) 5020 ml

18.4 200 measures

18.5 30 lengths

18.6 56 cm

18.7 5 g

19 Volume

19.1 (a) 88·7 cm^3 (b) 754 g
(c) 75·4 g (to 1 d.p.)

19.2 (a) 572 000 cm^3 (b) 572 litres

20 Enlargement and reduction

20.1 (a) *a* 27 mm, *b* 62 mm, scale factor 2·3
(b) 94 mm (c) 25 mm
((b) and (c) to nearest mm)
(d) 1·4

20.2 (a) 2·4 cm by 2·4 cm
(b) 1·6 cm by 1·6 cm
(c) Pupil's own 0·8 times reduction

20.3 (a) 3 (b) $\frac{1}{3}$ (c) $\frac{1}{4}$ (d) 2·5
(e) 1·6 (f) 0·625

21 Statistics

21.1 (a)

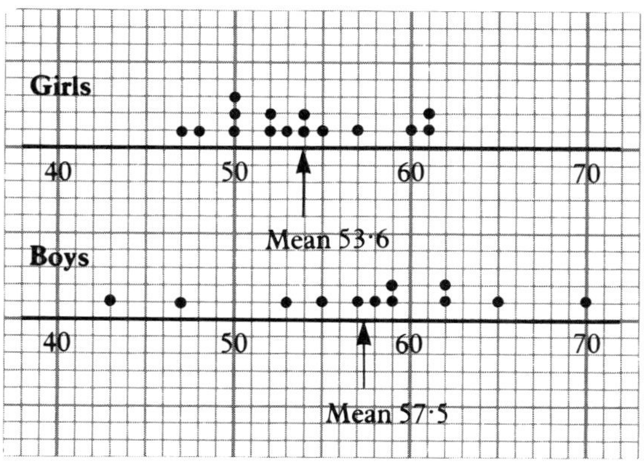

(b) Girls' mean 53·6 kg, boys' mean 57·5 kg
(c) The boys' weights are more widely spread out. Girls' range 14 kg, boys' range 27 kg

21.2 (a) 9·3 °C (to 1 d.p.) (b) 3·0 degrees
(c) No, it is impossible to find out the midnight temperature from this graph.

21.3 (a)

0–19	21%
20–39	27%
40–59	30%
60–79	17%
80–99	4%

(all to nearest 1%)
(b) 21%

21.4 (a) Mean pay $= \dfrac{260+220+120+100+4\times 90}{8}$
$= £132{\cdot}50$
(b) Six out of eight people get less than the mean wage.

22 Probability

22.1 (a) No, they both have the same chance of winning.
(b) $\frac{5}{200} = \frac{1}{40}$

22.2 (a) $\frac{3}{12} = \frac{1}{4}$ (b) $\frac{5}{12}$ (c) $\frac{4}{12} = \frac{1}{3}$
(d) $\frac{3}{12} = \frac{1}{4}$ (e) $\frac{2}{12} = \frac{1}{6}$

22.3 (a) $\frac{12}{18} = \frac{2}{3}$ (b) $\frac{8}{18} = \frac{4}{9}$ (c) $\frac{6}{18} = \frac{1}{3}$
(d) $\frac{10}{18} = \frac{5}{9}$

23 Arranging and selecting

23.1 (a) There are many possible solutions. Here is one:

	Mon.	Tues.	Wed.	Thur.	Fri.
A	Sci.		H.E.	Wood.	Met.
B	H.E.	Sci.	Met.		Wood.
C	Wood.	Met.		Sci.	H.E.
D	Met.	Wood.		H.E.	Sci.

Here is possibly an optimum solution for some schools:

	Mon.	Tues.	Wed.	Thurs.	Fri.
A	Sci.		H.E.	Wood.	Met.
B	Met.		Sci.	H.E.	Wood.
C	Wood.		Met.	Sci.	H.E.
D	H.E.		Wood.	Met.	Sci.

(b) It cannot be done, unless the rule about not having exams four days running is changed.

23.2 6 possible orders

23.3 9 ways

23.4 12 ways

M Miscellaneous questions

M1 £14·55

M2 2034

M3 Pupil's choice of number between 3·6 and 3·7

M4 £4·50

M5 About $1\frac{1}{2}$ hours

M6 0·35 litre

M7 £222

M8 14800

M9 (a) 0910 (b) 0830

M10 £20

M11 (a) 7 (b) 4

M12 Joan £1·80 per hour, Sadia £1·89 per hour
So Sadia was paid more.

M13 (a) August 5th (b) 0·170 kg (c) 170 g

M14 40 cm

M15 12 units

M16 (a) A ($^{-}2$, $^{-}2$), B (0, 2), C (6, $^{-}2$), D (4, 2)
(b) (2, 6)

M17 (a) Angle *c* (b) Angle *b*

M18 About 150 cm

M19 (a) 58·69 (b) 58·069, 58·099, 57·9

M20 (a) 186 units (b) £9·30

M21 (a) Box B (b) 100 bars

M22 (a) 15 boxes (b) 11 boxes

M23 (a) 576 (b) 96
(c) There were 4 nights of low attendance and 2 nights of high.

M24

(a) Either

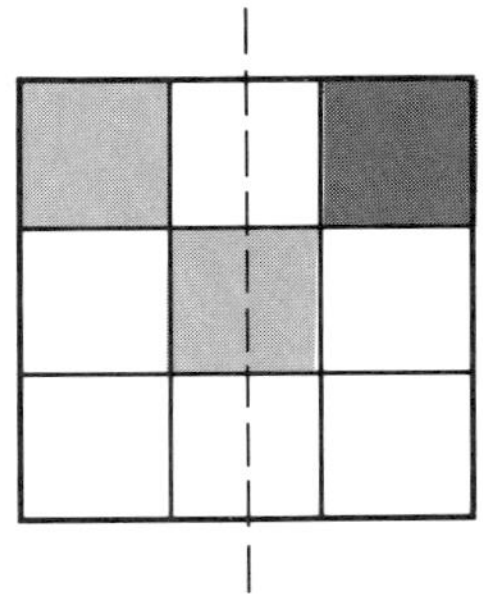

or

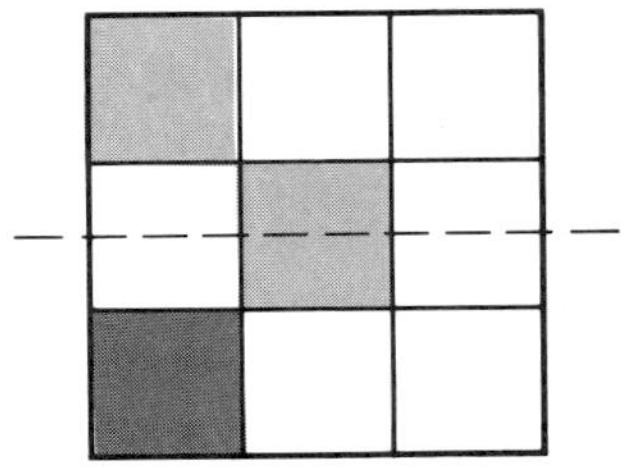

(b)

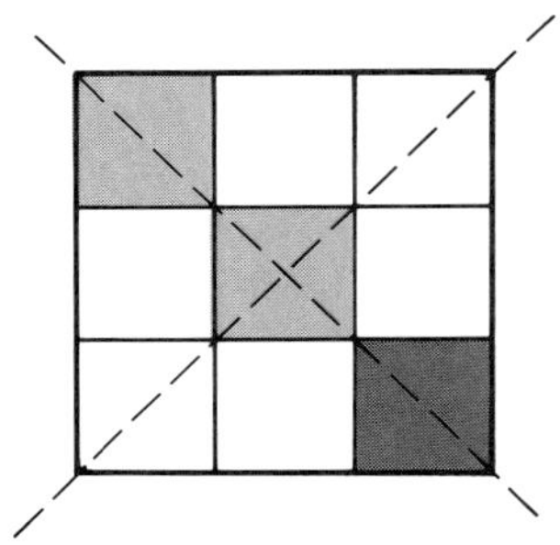

M25 (a) £234 (b) £238

M26 (a) R (b)

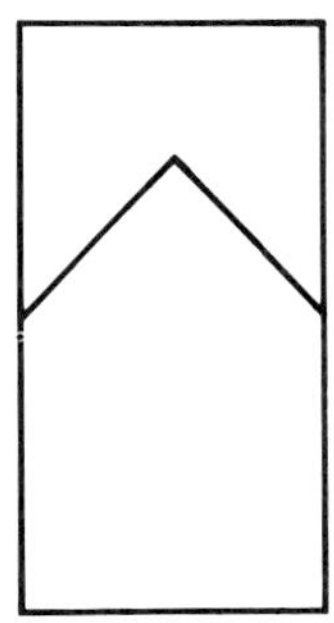

M27 (a) 148 pounds
(b) Pierre weighs about 147 pounds, so Martin is heavier.
(c) About 5 kg

M28 (a) A (0, 3), B (4, 1), C (2, $^-3$)
(b)

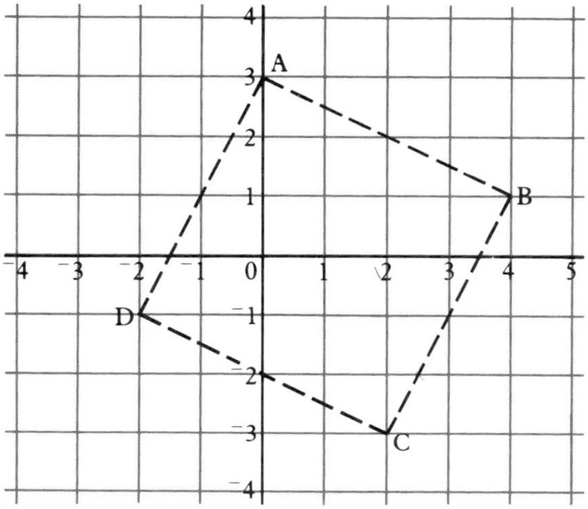

(c) ($^-2$, $^-1$)

M29 (a) 250 g (b) 125 g

M30 (a) Thursday (b) 2·4 litres
(c) 0·6 litre, 600 ml

M31 (a) p 8·4 m, q 17·6 m (b) 61·8 m
(c) 193 m^2 (to nearest m^2)

M32 (a) 1200 cm^2
(b) Pack A contains 30 000 cm^2.
Pack B contains 31 500 cm^2.
So pack B gives more paper for your money.

M33 CABED, CBAED, CEABD, CEBAD, DABEC, DBAEC, DEABC, DEBAC

M34 Blue, red, yellow, purple, green

M35 (a) No (b) No (c) Yes (d) No

M36 30 slices

M37 (a) £7·47
(b), (c) Two $1\frac{1}{2}$ litre bottles – £7·10
A 2 litre and a 1 litre – £7·04
(d) Buy a 2 litre bottle and a 1 litre bottle.

M38 (a) £3·90 (b) A, B, D or B, C, D

M39 89 days

M40 54 g

M41 (a) 18p (b) 12p (c) 125 cans

M42 (a) $\frac{3}{12} = \frac{1}{4}$ (b) $\frac{8}{12} = \frac{2}{3}$

M43 (a) 41·5 miles (b) 14 days

M44 (a)

Scale 1 cm to 10 m

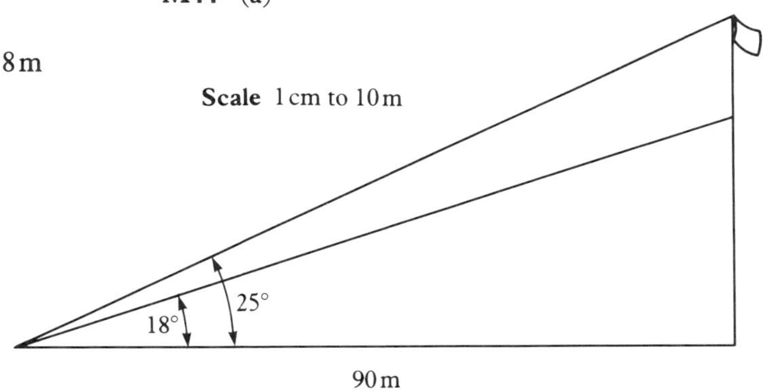

(b) About 13 m

M45 (a) (i) 125 calories (ii) 75 calories
(iii) 240 calories
(b) 440 calories

M46 (a) 720 m (b) 12 m/s
(c) 14·7 m (to 1 d.p.)

M47 (a) $x = 42°$, $y = 106°$ (b) 32°

M48

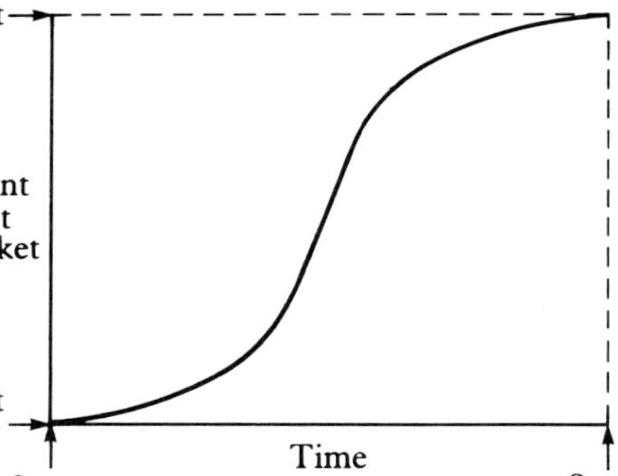

M49 (a) 65 cm (b) Waist 62 cm, shoulder 36 cm
(c) 80 cm

M50 (a) P yes, Q no, R no, S yes
(b) 272 cm^2

M51 (a) 0·723 52 m^3 (b) 724 litres

M52 Gillian's first-time passes 41·6%
Robert's first-time passes 44·8%
(both to 1 d.p.)
So Robert's driving school has the better record of first-time passes.

Published by the Press Syndicate of the University of Cambridge
The Pitt Building, Trumpington Street, Cambridge CB2 1RP
32 East 57th Street, New York, NY 10022, USA
10 Stamford Road, Oakleigh, Melbourne 3166, Australia

First published 1987
Reprinted 1988

Typesetting and diagrams by Marlborough Design
Printed in Great Britain at the University Press, Cambridge

ISBN 0 521 31465 8

ED